Soviet/Russian Laser Weapon Development

X-2 M & B-1 KLO/A-60 Ladoga ALK/Peresvet

HUGH HARKINS

Soviet/Russian Laser Weapon Development

X-2 M & B-1 KLO/A-60 Ladoga ALK/Peresvet

Centurion Publishing
United Kingdom

ISBN 10: 1-903630-81-9
ISBN 13: 978-1-903630-81-5

This volume first published in 2019

The publisher and author would like to thank all organisations and services for their assistance and contributions in the preparation of this volume:
JSC Concern Aerospace Defence 'Almaz-Antey' (JSC GSKB Almaz-Antey); PJSC 'Beriev Aircraft' Experimental Design Bureau; S.P. Korolev Rocket and Space Corporation Energia; FSUE 'GKNPTs them. M.V. Khrunichev; Ministry of Defence of the Russian Federation; Scientific and Production Association, S.A. Lavochkin (Laspace); Harkins, H. (2017) *Counter Space Defence Co-Orbital Satellite Fighter: The Soviet Istrebitel Sputnik Anti-Satellite Complex*, Centurion Publishing, UK; Harkins, H. (2016) *Iskander, Mobile Tactical Aero-Ballistic/Cruise Missile Complex* , Centurion Publishing, UK; Green, S.F. & Jones, M.H. (2015) *An Introduction to the Sun and Stars*, Second Edition, Cambridge University Press; National Museum of the United States Air Force; Central Intelligence Agency; Defence Intelligence Agency; Washington Post, 6 September 1985 (reproduced in CIA-RDP90-00965R000100120046-1, approved for release 2012-01-10); Kranaya Zvezda News

CONTENTS

INTRODUCTION

While there is a wealth of information available in regards to United States laser weapons program, including the application of such weapons in airborne laboratory testbeds, little is known about Soviet Union and later Russian Federation efforts to produce such weapons. This paper, which deals with the subject from a historical and scientific perspective, is intended to shed some light on the Soviet ground based KLO and A-60 Ladoga ALK laser weapon complex flying laboratory that first flew in the first half of the 1980's. The paper will also briefly detail Russian Federation continuation of A-60 ALK and KLO research work in the twenty first century, the former leading to the current generation Peresvet ground based combat laser complex unveiled in March 2018.

All of the technical data and the majority of the graphic material has been furnished by official sources within the current Russian and past Soviet defence industry as well as the Ministry of Defence of the Russian Federation and United States intelligence agencies.

1

FROM PARTICLE BEAM WEAPON THEORY TO DIRECTED ENERGY LASERS

In the modern world the laser (Light Amplification by Stimulation of Radiation), in its various forms, is taken for granted for many functions of civil society – medical procedures, manufacturing and astronomical research being but a few examples.[1] Investigations into potential military uses for laser technology began to emerge not long after the demonstration of the first practical laser system in 1960. The laser, which emits light waves in a single wavelength, stood out from other light sources that emit energy in a number of wavelengths – referred to as a spectrum (Green & Jones, 2015). While early military uses included employing lasers for range finding and targeting, the use of the laser as a hard kill weapon proved much more problematic for the scientific world despite their ease of integration into the world of sci-fi in many forms such as the venerable light saber of the 'Star Wars' movie franchise – in the real world of science the light saber is a practical impossibility. Even in the second decade of the twenty first century the laser is showing less promise as a hard kill weapon than it does in a defensive application – laser based opto-electronic countermeasures systems designed to decoy an incoming ground launched or air launched IR/IIR (Infrared/Infrared/Imaging Infrared) guided missiles for example. In this respect the laser can also be used in an offensive capacity by disabling systems onboard a surface, airborne or space-based platform.

The limitations of the laser as a hard kill weapon was not lost on science in the last few decades of the twentieth century, there being other, potentially more promising and more powerful areas to explore in the race for an ultimate weapon system – a Sword of Damocles if you will – to hang over the head of an enemy.

PARTICLE BEAM WEAPONS – While laser technology was finding applications in many areas, including military roles such as range-finding, the very high power/fuel requirements for any hard kill application saw attention focused on

---

[1] The laser produces a beam of thermal radiation in a line of advance from its origin to the point of absorption

technological research in the futuristic sounding field of particle beam weapons. By the early 1980's there were growing cries of doom and gloom about Soviet supremacy in laser and particle beam weapons development, building on research gleaned through many processes from theoretical to fundamental applied physics. The first Soviet Union (USSR – Union of Soviet Socialist Republics) particle beam accelerator was apparently built near Moscow from the mid-1950's, but this did not become fully operational until well into the 1960's. Its scientific research role was not necessarily of a purely, or indeed, a primary military nature. In point of fact, it would be a fair assessment to conclude that military research was a byproduct of the wider civil scientific research programs then ongoing in the Soviet Union. Much of the United States intelligence claims about Soviet directed energy advances were wildly exaggerated in order to exhort public support, which could translate into funding, for the hugely expensive and ultimately, considering technology levels available in the 1980's, technologically unachievable SDI (Space Defense Initiative). The SDI – referred to as 'Star Wars' after the 1977 blockbuster movie – and Soviet planned countermoves amounted to wholesale plans for the militarisation of near Earth space, a major threat to political and military stability between the two power blocks back on Earth – NATO (North Atlantic Treaty Organisation) and the Warsaw Pact.[2]

The Russian Peresvet mobile ground based combat laser complex, when unveiled in March 2018, was something of an enigma in regards to role and capability. MODRF

---

[2] The Warsaw Pact was a formal treaty of friendship, co-operation and mutual assistance signed between the Socialist Republics of the USSR and 7 Soviet Orbit satellite states in Eastern Europe. This treaty, which took effect from 14 May 1955, was designed to counter the growing NATO alliance, formed on 4 April 1949, opposed to the Eastern Block dominated by the Soviet Union

The overly suspicious, at times surpassing the 'Reds under the Beds' paranoia, attitude inherent within the western intelligence agencies in the 1970's brought about conclusions, based completely on hearsay and conjecture rather than scientific evidence, that the Soviet orbital space stations of the Salyut series were part of an elaborate space weapons program to deploy laser and particle beam weapons in orbit.[3] An example of this paranoia can be interpreted from the following statement by General Dan Graham of the American Security Council, for a science documentary in the early 1980's: the Soviets had '45 men in space in the six year hiatus between our [United States] last moonshot... [Apollo 17, mission ended on 19 December 1972] and the shuttle [this period covered 20 December 1972-12 April 1981] and they weren't up there trying to set records... they were up there for serious reasons'. Of course, the Soviets were in space for serious reasons, none of which, was there evidence to support, being related to particle beam and laser weapons programs as was the implication of the above statement that certainly appeared designed to leave the audience to infer that the Soviets were up to some skullduggery in space. By the same mentality it would have to be inferred that the current International Space Station in Earth orbit was the venue of a grand conspiracy to dominate us mere mortals back here on Earth, whereas in actuality the only weapons on board (actually stored on Soyuz spacecraft docked with the ISS) are special design pistols designed as emergency hunting weapons for crews returning to Earth in an emergency and landing well off course in some hard to reach part of the planet.

In the United States, particle beam weapon technology was researched by the Lawrence Livermore Laboratories in California. This initially focused on the idea of an experimental complex known as an electron gun, tested to the power of 3 million eV (Electron-volts.[4] Negatively charged electrons repel each other requiring them to be forced towards each other through coils, following which they are transferred in accelerator modules where the energy is increased to 5 million eV. The theory was that once fired the particle beam would be held together through a magnetic field that was generated around it. Dr. Richard Briggs of Lawrence Livermore Laboratories stated:

> "A beam which is held together by its own magnetic field in an air media like this [the particle beam accelerator] can undergo kink like disturbances and begin to thrash about inside the tube. In fact under some conditions we find that the beam hits the side of the tube and can damage it..."

The damage to the tubes was merely cracks in the glass. However, at the power output of 5 million eV there was insufficient output for the magnetic field to hold the electrons together. Such research confirmed the need for much higher power

---

[3] Salyut was active from 19 April-11 October 1971 (1 crew for 22 days); Salyut-4, 26 December 1974-3 February 1977 (2 crews totaling 28 and 63 days); Salyut-6, 29 September 1977-29 July 1982 (5 prime and 11 short duration crews) and Salyut-7, 19 April 1982-7 February 1991 (4 prime and 5 short duration crews) (Energia). In the 1980's, a battle station complex based on the DOS-7K orbital station was studied. This would be designed to engage Earth surface targets

[4] 1 eV = 1.602 x 10^{-19} J (Joules)

output from next generation accelerators, Livermore then looking to energise the electrons to an output of 50 million volts. What had become clear was the impracticality of particle beam weapons, as to produce the power output of 50 million volts required an accelerator the size of a medium to large size factory building. To produce levels of power required to inflict large scale damage to a target would have, if such weapons were feasible, required an accelerator the size of a city, hardly practical for a system to be installed in a ground vehicle, airborne platform, orbiting spacecraft, or even, as was the first intended application for a such a weapon, a converted aircraft carrier naval combatant.

While it is clear that particle beams would be more efficient in a space environment bereft of the major effects of the Earth's atmosphere, this would not be on a scale that would allow such a weapon complex to be significantly reduced in size to allow it to be deployed in an orbiting spacecraft. Of course, there were other potential offsets to such theorised weapons as particle beam weapons, not least of which was the school of thought, born out by mathematical data available in the early 1980's, that with the equal and opposite charges that would inevitably built up in exit of the accelerator, then the particle beam could itself be dragged back into the accelerator tube with catastrophic consequences. The most serious problem, perhaps, even if the problem of the particle beam being drawn back into the accelerator tube was overcome, was the conclusive fact that the electrons in the beam would repel each other in the uncontrolled non-laboratory conditions of Earth orbit. This would have caused the beam to flay-around with no real hope of projecting it onto a concentrated point – in other words targeting would not be possible. Even if the former problem was overcome, targeting would be seriously affected by the effects of the geomagnetic field of the Earth, which would cause significant bending in the beam, making it impossible to target a fixed location let alone a missile in any stage of its flight – supersonic or hypersonic.[5] Distance, of course, was another factor in that even if the beam could be brought to bear on a target it would have to be at close range as the effects of the beam massively degraded with increasing distance to the point that by several hundred miles it would have been be ambitious to expect anything more than blistering paintwork on a space vehicle body.

Many programs were implemented, some possibly still in small-scale existence today, to overcome the problems associated with particle-beam weapons. One such American program, which researched the area of overcoming or avoiding the problems associated with development of particle beam weapons, was set up at Los Alamos National Laboratory, New Mexico. This program was aimed at developing a neutral beam using hydrogen atoms, which have the advantage of having no charge – neutral charge. This was looked upon as mission impossible as an accelerator required the charge in order to energise the particles. Neutral particles, it was clear would not be visible to the electromagnetic field used in the energy transfer process. To overcome this it was decided to attach charged electrons to the neutral hydrogen atoms, which would then be accelerated and the electrons removed from the

---

[5] Supersonic speeds range from Mach 1 to Mach 5 where hypersonic speeds begin

hydrogen atoms. In regards to this research a Dr, Ed Knapp, head of Acceleration Division Los Alamos National Laboratory stated:

"After the negative hydrogen ion beam comes the ion source and this confined and focused and prepared, it enters the first new piece of accelerator apparatus [assumedly developed by Los Alamos] …a device called the radio frequency quadruple. There are four poles located as such, forming a channel in which the particle beam can move. These four poles are charged, producing an electric field which holds the particle beam very tightly in the absence".

This work was based on technology incorporated from a Soviet design apparently developed at the Novosibirsk laboratory, which, as admitted by Dr. Knapp, was 'copied' by the Americans from open source scientific literature. Despite the paranoia in the west that the Soviets were developing a futuristic weapon based on particle beam technology at a greatly accelerated pace, the nature of much of the particle physics research being conducted, as noted above, was open source, as evinced by the fact that the Los Alamos team were able to base their work on that of Soviet physicists. However, the fact that western advances in particle beam development was based on Soviet technology added ammunition to those scaremongering voices shouting out about the Soviets leaping ahead in advanced space-based weapons technology. In regards to the possibility of this technology being able to produce a particle beam weapon system, Benjamin Sidhorov, Deputy Director of the Soviet Institute of Nuclear Physics, Novosibirsk, in the early 1980's stated that, using then available technology, a particle beam weapon was possible from a physical or theoretical point of view, but not from a technical point of view. It later became clear from interviews that the Los Alamos team was under no illusion that the research that they had embarked upon simply was not able to produce a viable weapon system using then current and projected technology levels; the Soviets appearing to have arrived at similar conclusions earlier.

LASER – The laser was invented in the early 1960's, physicists Alexander Prokhorov, Nikolay Basov (Soviet Union) and Charles Townes (United States) jointly receiving the Nobel Prize in 1964. Regardless of intention, as happens with all great scientific advances, studies were conducted to pursue the possibility of military applications for the new creation. With the advancement of laser technology through the 1960's and 1970's, the science's potential as a weapon was a powerful magnet for both power blocks of East and West in their drive to achieve superiority in weapons technology.

In the late 1970's and into the 1980's it was clear that, with then current and projected near to medium term technology levels, high powered laser systems, despite their drawbacks, held more promise for development as a viable directed energy weapon system than particle beam accelerators. The laser as a weapon certainly appeared to be more scientifically attainable. The basics of a carbon dioxide laser facilitates for the gas to be forced into a chamber where it is excited by exposure to an electric discharge. This allows the emission of photons of light that are narrowed into a beam by combinations of mirrors. The carbon dioxide laser beam, present in the infrared spectrum, is actually invisible to the naked eye, the characteristic pinkish/reddish hue associated with such beams being the product of gasses.

A viable laser weapon system would negate the need for heavy fuel hungry and expensive to manufacture missiles – missile body, rocket or jet engines and boosters, individual guidance systems and, of course, the warhead. Instantaneous – speed of light[6] – target impact as opposed to a missile flight time that could be up to several tens of seconds. In regard to an airborne platform specific laser weapon complex there was the potential for striking targets at ranges significantly in excess of the range of air to air missiles and no need for high carrier aircraft speeds and acceleration to gain launch energy as is required for an optimum missile shot. While a Mach 2 capable launch platform is able to impart more energy on its missile at launch than a Mach 1 capable launch platform and is therefore more desirable from an operational standpoint, the speed of the laser weapon complex carrier platform is of new consequence to the energy carried in the thermal radiation beam, which, unlike the homing head of a missile, cannot be jammed.

There were of course many problems that would have to be overcome if the laser was to be developed into a viable weapons system. While demonstrations as early as the 1970's and into the 1980's showed that if focused on a target long enough a carbon dioxide laser could inflict damage on a target in laboratory and controlled test environments, the practically for use as a weapons system in a real-world environment was clearly absent. To achieve even a rudimentary success in tests required immense power supply taking up a large volume for a system that would only be effective at very close range – downwards of a few hundred meters at best. Even these limited capabilities would have required almost perfect clear and calm atmospheric/weather conditions. Cloud, fog, rain, mist and wind, or lack thereof, and even smoke, could severely impair the laser weapons capability to the point of ineffectiveness. As an example, the standard guided bomb unit through the latter decades of the twentieth century and continuing through the second decade of the twenty first century is the so called LGB (Laser Guided Bomb). Such weapons are effective only in reasonably fair weather conditions. A heavy overcast completely negates their effectiveness, which can also be adversely affected by precipitation, fog and smoke. In fact, the LGB has such severe drawbacks that with the advancing of GPS (Global Positioning System guidance units for bombs, the latter is coupled with the laser guidance unit on such weapons as the Raytheon Paveway IV. The laser guidance, assuming a competent operator and favourable operating conditions, is more accurate whilst the GPS system can be employed when cloud cover is present over a target area. In regard to wind speed, higher winds are actually beneficial as it sufficiently prevents the laser beam overly heating the air, thereby reducing the distortion of the laser beam by hot air.

In a real world environment an actively defensive avoiding airborne target – this was among the main target sets envisioned for laser weapons – flying at high – Mach 1-1.5 – speed, a laser weapon available in the 1980' would have had doubtful

---

[6] Speed of light in a vacuum, represented by the symbol c, is 3.00×10^8 m s^{-1}

effectiveness in its ability to be accurately brought to bear for several seconds. The best counter to laser weapons was the atmosphere itself, particularly at lower altitudes were the air is densest and Earths weather mechanisms are very active[7] – this resulting in a phenomena that is referred to as thermal blooming. Thermal blooming (also referred to as thermal defocusing) is the reduction of irradiation levels of a laser beam in its transit through Earth's atmosphere. Experimental models to determine the effects of atmospheric thermal blooming on CO_2 laser have been run at various institutions. In short, results show energy of the light beam is lost through a process of absorption of molecules in the atmosphere and atmospheric scattering of molecules and aerosols. High power lasers of the type required to produce destructive damage to an aircraft or missile in flight results in atmospheric heating along the line of the beam as a product of the above mentioned molecular absorption. This leads to processes that ultimately reduces the power of irradiation that was initially intended to neutralise the target – thermal blooming. Adverse atmospheric conditions – cloud and precipitation – will further degrade the irradiation energy through increased thermal blooming.

The problems of thermal blooming have not been overcome despite the small scale deployment of limited capability laser weapon complexes in the second decade of the twenty first century. Such weapons have been deployed as very short range point defence systems on large warships by necessity as much space is required for the various components and fuel. While demonstrations of such weapons have shown limited promise in hotter climes more akin to clear weather conditions, they would be severely constrained in a North European or Atlantic geographic location where they would be particularly vulnerable to thermal blooming. The laser deployed in a trials capacity by the United States Navy is effective only against small slow flying drone style targets, although it is stated that the type also has some potential against RPG (Rocket Propelled Grenades) man portable anti-tank weapons in controlled environmental conditions.[8] Such laser weapons would be ineffective against such high performance weapon systems as the Russian Yakhont anti-ship cruise missile capable of high supersonic speeds. They would also be of little use against subsonic anti-ship cruise missiles flying 5-10 or so metres above the sea surface, moisture and spray reducing the lasers thermal energy to the point of

---

[7] Earth's atmosphere is made up of six major layers – the Troposphere (begins at the surface of the planet and extends to an altitude of 8-14.5 km) is the densest layer containing most of Earths weather mechanisms; the Stratosphere, which contains the Ozone layer that protects the planet by absorbing and scattering solar ultra-violet radiation, takes over where the Troposphere ends and extends to 50 km altitude; the Mesosphere begins where the stratosphere ends and extends to an altitude of 85 km where the Thermosphere begins, this extending to an altitude of 600 km (the orbital region of many satellite, including the International Space Station. The Ionosphere, which overlaps the Mesosphere and thermosphere regions, consists of a layer of electrons, ionised atoms and molecules – a highly dynamic layer that can extend to altitudes of 965 km, but can shrink to considerably lower altitudes depending on solar conditions prevailing. The outer layer of Earth's atmosphere is the Exosphere, which can extend to altitudes of 10000 km (NASA Goddard)

[8] Such projectiles can reach velocities of several hundred over a very short range m s^{-1}

ineffectiveness – sea spray would also induce severe thermal blooming against an RPG target which would inevitably follow a more or less sea level flight trajectory.

Boeing NKC-135 Airborne Laser Flying Laboratory during a flight over the New Mexico Desert. The NKC-135, like the Soviet A-60 Ladoga, conducted ground breaking research into weapons application for the CO_2 laser, neither leading to a viable weapon system. NMUSAF

As was the case in the Soviet Union the United States Department of Defence had an active laser weapon research program in the 1970's and 1980's. Initially this was designed for airborne application in a modified Boeing KC-135 Stratotanker, which was, post modification, having a CO_2 laser installed, referred to as the NKC-135. This concept was employed in a test program in which the goal was the destruction of an air to air missile in flight. A test firing of a laser against an AIM-9 Sidewinder short-range air to air missile in summer 1982 ended in failure. A further test the following year met with success, the target being destroyed. However, a rudimentary success in perfect conditions at relatively short-range against a target on a predictable easy to track path belied the impracticality of the laser as an effective weapon system in the early 1980's, particularly considering the rise to prominence of the ambitious Regan era SDI (Space Defense Initiative) program. This envisioned such weapons being placed in orbit to defend the United States against ballistic missile attack – the simplified theory of the system being to destroy the ICBM

(Intercontinental Ballistic Missile) warheads in low-Earth orbit before they reentered the atmosphere in the terminal phase of the flight to the target. Again, the problems to be overcome were many and, with the technology levels then available, insurmountable in the short-term to medium. Even if deployed in a low-Earth orbit space environment where atmospheric interference would be negligible[9], the spacecraft mounted mirrors required to achieve a range of ~1600 km (the operational range would have had to have been considerably in excess of 1600 km), was estimated to be in the order of 4 meters in diameter, eight to ten times larger than the largest mirrors then in use. Even after producing these large mirrors there remained the problem of placing a sufficient power source into orbit, requiring something the size of a four engine airliner for the equivalent of one or two shots, something completely impractical, even at 2019 technology levels. In the early 1980's a four meter mirror was simply not practical. Even if practicality was overcome, such a mirror would have cost perhaps a billion dollars (around US $2.57 billion in 2017 values when inflation is added) or more based on the then projected costs of around 400 million dollars for a 2.8 m class mirror.

In the early 1980's, analysis of satellite imagery led to a CIA (Central Intelligence Agency) conclusion that a Soviet laser test centre was located at Krasnoarmeysk, around 30 miles to the North East of Moscow. However, although laser research and test centres existed in the Soviet Union poor intelligence data led to alarmist reports, in September of 1985 to the effect that the Soviets were serially producing laser weapon complexes. At this time it was reported that the Soviets employed a 'magnetohydrodynamic (MKD) [MHW] generator, which produces 15 megawatts of short-term power' (WP, 1985, citing CIA data). The text of a US State Department report, classified as secret and reproduced in the Washington Post read 'it [the MHW] generates current by passing a conducting fluid through a magnetic field'. The report continued:

> "Their [USSR] MHD work is the largest in the world and continues to grow. Power outputs already achieved exceed those in the West several fold, and both rocket power and liquid metal system inputs could have potential for military programs in high-energy lasers, charged particle beams and space-borne laser power supplies. MHD is a technology area where the Soviets clearly lead the U.S in demonstrated capability". (CIA-RDP90-00965R000100120046-1 – Approved for release 2012-01-10).

While the evidence that has emerged over the ensuing three plus decades shows that the Soviet Union had made enormous breakthroughs in laser weapon development, the US intelligence reports of the mid-1980's were wildly exaggerated, in part to illicit funds for the SDI program. There were several Soviet development programs, but none of these were at that time ready for serial production and service introduction.

---

[9] As shown in footnote 7 the outer layer of Earth's atmosphere extends to an altitude of ~10000 km

In the 1970's the new generations of NATO and Soviet tactical strike aircraft, typified by the variable-geometry Panavia Tornado GR.1 (top) – RAF aircraft in September 1995 – and the Sukhoi Su-24M (above) – Russian Aerospace Force aircraft operating in Syria in 2015-2016 – were equipped with laser technology to assist in targeting. The Tornado was equipped with a laser range finder and the Su-24M was equipped with a system referred to as the LTVSTS (Laser/TV Search and Track System). In the Syrian campaign of 2015-17 the Su-24M fleet launched a limited number of Kh-25ML laser guided air to surface missiles. Author/Sukhoi

2

TWENTIETH CENTURY KLO & ALK - TWENTY FIRST CENTURY PERESVET

Most Soviet developments with laser (Light Amplification by Stimulation of Radiation) technology as a hard kill weapon remain classified in the main successor state, the Russian Federation. However, enough information has come to light in the second decade of the twenty first century to allow us to reach an informed view of the major KLO (Complex of Laser Weapon) ground based platforms and the ALK (Airborne Laser Complex) that spawned the Almaz-Antey/Beriev A-60 Ladoga LKAB (Laser System Air Based) – the Lazernoye Oruzhiye airborne laser experimental laboratory. We know that there was a substantial development effort in the area of laser weapon complexes in the Soviet Union in the last decade and a half of the Cold War period – mid 1970's through 1991 – but that through much of the 1990's such programs fell into more or less a state of suspended animation following the dissolution of the Soviet Union on 25 December 1991.

Whilst the Soviet Union was advancing laser technology primarily as a defensive capacity weapon(s) NATO was focused on the theoretical Soviet drive toward the laser as a weapon of attack. Defensive or offensive, the theory of the laser weapon seemed attractive for a number of reasons. A target could be struck almost instantaneously whereas a missile could take several tens of seconds and interception could be effected at longer range as noted above. Defensive avoidance maneuvering of the target would not be an issue with the use of the laser, assuming sufficient energy was focused on said target to effect destruction in the first second or so. Each laser shot was expected to cost a small fraction of the cost of a surface to air missile. The accomplishment of the mission objective could be achieved through either hard kill (destroying the target) or soft kill (disabling opto-electronic systems). Removal of the requirement for the long-term accumulation and storage of ordnance such as surface to air missiles located in an armoury (of course fuel for the laser complex would have to be stockpiled). When employed against targets at or near ground level then high accuracy of the laser would remove much of the threat of damage to the local environment in the vicinity of the target.

The potential for the successful operation of an ALK, which would emit energy in wavelengths that could be 'strongly absorbed in the atmosphere' (Almaz-Antey), was always going to be greater when employed at higher altitudes against targets themselves flying at altitudes above the dense lower layer of the atmosphere – the Troposphere where Earth weather systems are active – this could be against Stratospheric or exo-atmospheric[10] targets (the longer the range of shot the weaker the energy focused on the target would be). When mounted on an aircraft platform the ALK would be able to rapidly move from one geographical locality to another. Of course, such a weapon complex was not going to be accommodated in a tactical combat size aircraft. The large size carrier aircraft – necessitated by the need to accommodate the various components and fuel of the laser complex – had the knock on effect of providing an airborne platform with extended range, allowing a potential operational ALK system to remain on combat loiter for considerable periods during times of tension rather than remaining on ground alert where it was vulnerable to a pre-emptive strike.

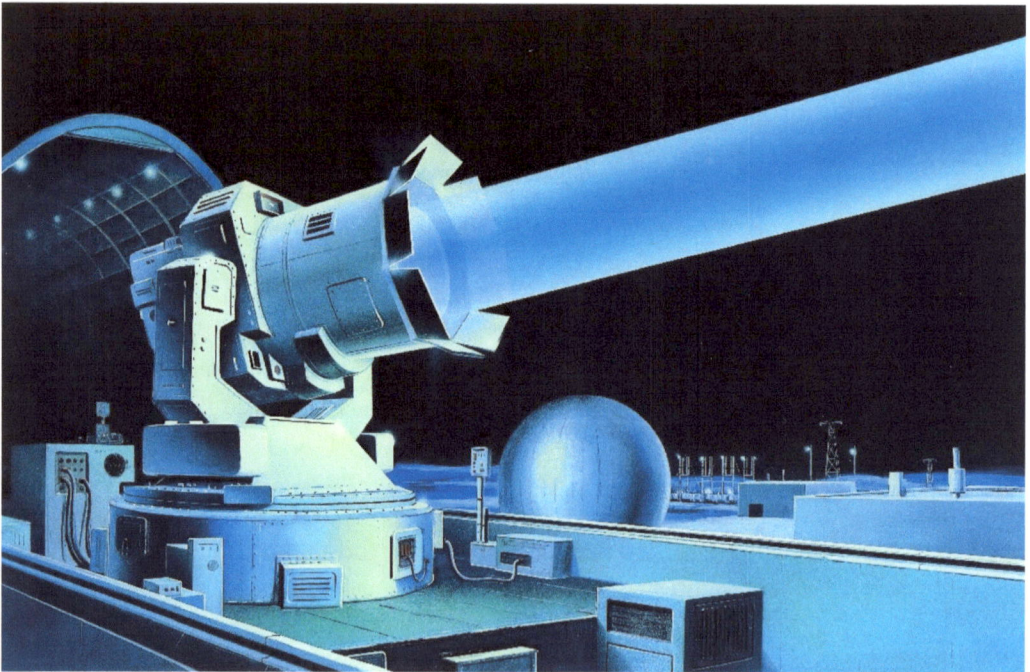

A 1986 DIA depiction of an intelligence assessment of a ground based laser weapon test stand, presumably at the Sary Shagan test facility. DIA

The origins of the Soviet laser weapon program that spawned the KLO surface based laser complexes and the ALK A-60 Ladoga can be traced back to June 1965

[10] The term exo-atmospheric here represents orbital and sub-orbital regions where a space vehicle would orbit the Earth while still being within thin outer layers of the atmosphere – Ionosphere and Exosphere

when a conference was held attended by academicians A.M. Prokhorov and M.D. Millionshchikov. That the main role envisioned for a future laser weapon complex was air and possibly space defence is evinced by the fact that a team of specialists in the area of surface to air missile technology was formed within OKB (Almaz) under the supervision of B.V. Bunkin. This research group was added to by specialists recruited from various universities. In August 1973, this research group was used to form the basis of a special design bureau known as NGO (CDB) Diamond[11], which had the explicit purpose of furthering the development of the laser as a practical weapon system. The new design bureau was headed by chief designer (later Doctor of Technical Sciences) Theodore Brahman. From 1975 the design bureau was headed by Doctor of Technical Sciences Lev Zakharov. CDB Diamond was added to by personnel from other departments and established cooperative links with various institutes and production centres under the control of Almaz (Almaz-Antey).

A 1987 DIA depiction of a mobile ground based anti-air warfare laser weapon complex defending a Soviet air base against air attack by NATO General Dynamics (now Lockheed Martin) F-16 tactical strike fighter aircraft. This depiction appears to have been based on data acquired about one of the Soviet mobile ground based laser weapon complexes tested in 1982. DIA

CDB Diamond was a world leader in air to surface missile capability having developed the S-25, S-75 and S-200 air defence missile complexes. The organisation

---

[11] CDB Diamond is a part of the modern day JSC GSKB Almaz-Antey

was, in the 1970's, advanced in development of the S-300P long-range air defence missile complex that would lead to the modern day S-300 series that constitute a major component of the Russian Federation air defence capability alongside the S-400 Triumph long-range air defence missile complex.

Soviet experimental KLO – Complex of Laser Weapons – ground based mobile laser weapon complex. The complex was carried on a development of the Maz-543M eight wheeled chassis. The laser complex was located toward the forward part of the carrier and was raised when ready for firing operations. Almaz-Antey

In the period covering the late 1970's and early 1980's, considerable research was conducted into the application of laser weapon technology, including the production and utilisation of a number of test stands for testing the laser complex. In this phase of the program CDB Diamond developed the necessary components to enable the generation and targeting of a powerful beam of radiation against targets that were described as 'remote' (Almaz-Antey), but with no elaboration on specific range(s).

The research conducted in the late 1970's and into the 1980's led to development of two related programs – the ground based complexes and an airborne complex, both able to engage and destroy targets through the process of thermal overload. For this it was necessary to project a 'narrow' beam of thermal radiation that could be densely concentrated on a specific target point at considerable distance from the laser complex (Almaz-Antey). Through research conducted by groups under the overall scientific supervision of Academician A. Prokhorov, B.N. Bunkin and E. Velikhov, a research team headed by Professor Leonid Zakharov, Academician of the USSR (Union of Soviet Socialist Republics) Academy of Sciences of the Deaf, and other research centres, the first practical examples of laser weapon complexes were developed. Two distinct types of system, able to destroy or damage material objects through the process of thermal overload, were ground tested under laboratory and full-scale operation conditions. These complexes, designated X-2 M and B-1[12], were extensively tested against air targets under what was termed 'full-scale conditions' (Almaz-Antey).

Close up of the telescope mounting of the laser weapon firing head for a Soviet experimental KLO ground based laser weapon complex. Almaz-Antey

The basic composition of the ground based KLO consisted of a mobile complex – based on the MAZ-543M chassis[13] – for carriage of the laser weapon itself; a data collection and thermal radiation beam aiming system; a system for guiding the thermal radiation beam and retaining same on the target, all conducted through the operation of an automatic control system.

---

[12] One of the ground KLO based complexes had the stenciling ПДС (PDS)

[13] This chassis had a load capacity of 22000 kg, engine power of 526 PS and a maximum road speed of 60 km/h

"The output elements of the transportation system and the formation of powerful radiation, the receiving elements of the information-aiming system, the executive elements of the guidance system and the retention of powerful radiation on the target are combined into a single WMD (Optical-Mechanical Device)" (Almaz-Antey).

In development of the KLO ground complexes a high-precision optical sighting system, referred to as an 'optical information sighting' system, was developed (Almaz-Antey). This system, which received the various data on the specific target to allow for accurate targeting, was required to be shielded from the interference of the optics due to high levels of radiation energy impacting the target (Almaz-Antey).

Lavochkin La-17K jet powered uninhabited target aircraft of the type employed in the Soviet laser weapon tests. Laspace

A typical target engagement sequence would follow the following format: "According to preliminary designation information, WMD [Optical-Mechanical Device] is deployed in the direction of the target. The information-aim system detects the target. Information is transmitted to the executive elements of the guidance system, which realize automatic tracking of the target and alignment of the optical axis of the system with the direction to the target. At a given time, a powerful laser begins to generate radiation. Through the system of mirrors, the radiation enters the output telescope with a large-sized reflector, which forms it into a narrow beam in the direction of the optical axis of the system. The guidance system ensures stabilization of the beam on the target during the entire period of power transmission. The BVO guidance system quickly changes the orientation of the powerful beam in space" (Alamz-Antey).

The basic KLO ground complex had been developed and built by 1980. A series of system testing of the guidance units and the generation of the thermal radiation beam had been successfully developed. Such systems development was ahead of development of the actual laser device resulting in a delay in full-scale testing until summer 1982. By the middle of September that year the system was ready and Y.A. Konyaev, overseeing the technical aspects of the testing, filed a report to L.N. Zahareva, chief designer of the complex, indicating that the system was ready to commence a live fire test against an airborne target. Preparations for the test continued through the night of 21-22 September, all being ready by 05.00 am on the morning of the 22nd. In the test site locality the weather conditions – dry and favourable wind conditions – were more or less perfect for the test shot and the go order was given for the target aircraft to be launched. This target aircraft has been speculated to be a Lavochkin La-17r uninhabited jet powered monoplane. However, the La-17r was a reconnaissance variant of the La-17 (index 201) target drone family. The most likely variant was the La-17K as this was the standard production variant at that time – serial production commencing in 1978 and continuing into 1993. This small size aircraft was powered by a single R-11F2S-300 (in the La-17K this modification was designated R-11K-300) turbojet engine mounted on the extended lower centre fuselage section (Laspace).

The Beriev/Almaz-Antey A-60 Ladoga ALK laser weapon flying laboratory 1A1. Almaz-Antey

Although targeting was apparently successfully conducted on the L-17K air targets first pass over the fire zone, the technical controller made the decision to conduct the thermal radiation shot on the targets second pass. The order was given to fire and, after a very short period, the thermal heating overload on the target resulted in a flash on a wing, which was then noted to fall away from the aircraft, which then went out of control and crashed to the ground. From the firing of the laser to the flash that removed the wing from the target fuselage only a few seconds had elapsed (Almaz-Antey).

The Beriev A-60 Ladoga developed as the LKAB platform for the Soviet airborne laser complex was a development of the Beriev A-50 AWACS (top), which was itself developed from the Ilyushin Il-76 transport aircraft (above). MODRF/Ilyushin

Following the successful test and the obligatory honours for the lead figures of the many thousands involved across many bureau, departments and institutes, attention turned increasingly to the implementation of a similar test program for the ALK LKAB program, work on which had commenced in the late 1970's. L.N. Zahareva team, which brought much experience from the ground based laser complex programs, worked under the overall supervision of B.V. Bunkin on the ALK program.

The advantages of the ALK over the KLO were its intended operation at upper altitudes were the absence of the weather associated with lower altitudes translated to vastly reduced possibilities for interference of the thermal radiation beams path to the target. Operating above the highly dynamic lower layer of the Earth's atmosphere provides better conditions for reducing, even removing, thermal blooming. However, there were also disadvantages in operating from an airborne platform. Not least among these was the fact that the laser would have to be precisely targeted on a fast moving target when itself mounted on a fast moving platform. In addition, while it is possible to provide a stable launch platform on a ground based complex this was fraught with difficulties on a large airborne platform which would be prone to vibrations due to a number of factors, not least being the powerful engines powering the aircraft. There would also be significant differences in the outside temperature of an aircraft at perhaps 8000 m compared with sea level temperature – this operating environment requiring modifications to the equipment and sensors.[14]

Taganrog Aviation Scientific-Technical Complex – G.M. Beriev – was brought into the program as the LL (Flying Laboratory) aircraft platform developer under chief designer A.K. Konstantinov (Almaz-Antey & Beriev). The platform selected as the LL equipped with the LKAB was a development of the Beriev A-50 AWACS (Airborne Warning and Control System), which was itself a development of the Ilyushin Il-76 four engine jet powered transport aircraft.[15] The new A-60 design, the first example of which was allocated the index 1A1 and the name Ladoga (Almaz-Antey), lacked provision for the specialised equipment required for the AWACS role and introduced a plethora of new equipment associated with directed energy experimentation. In general appearance the most notable differences between the A-60 and the A-50, other than the obvious lack radar rotodome atop the fuselage of the former, was the enlarged dome housing radio-electronic equipment on the nose of the A-60 and the more pronounced lateral fuselage body bulges. The overall livery remained the same as that of the A-50 and some Il-76 transport aircraft. Early photographs show the A-60 1A1 with Aeroflot stenciling on the forward fuselage.

As the ALK LL A-60 1A1 conducted its maiden flight on 19 August 1981 (Almaz-Antey). The aircraft was based near Moscow – Chkalovsky (Shchelkovo) military airport. Meanwhile, work continued on development of the LKAB to be carried by the A-60. A large number of ground simulations were conducted to reach

---

[14] Although there are differences in model calculations, a ~6.5 deg. C fall in temperature can be expected for every 1000 m altitude increase in the Troposphere. In the stratosphere the temperature is coldest at the bottom of the layer and warmer at the top of the layer

[15] The A-50 conducted its maiden flight in December 1978. The Il-76 prototype powered by D30-KP engines conducted its maiden flight on 25 March 1971 and was accepted into service with the Soviet Air Force in 1974 (Ilyushin)

a preferred combination of structural and systems satisfaction before the installation of the laser complex in A-60 1A1 (Almaz-Antey).

Specification for the Beriev A-50 from which the A-60 was derived – A.M. Beriev

Maximum take-off weight: 190000 kg
Flight endurance during mission at 1000 km range
    Without in-flight refuelling: 4 hours
    With one in-flight refuelling: 7 hours
Flight crew: 5
Systems crew: Unknown

A-60 Ladoga 1A1 with aircrew and program personnel. It is unclear if this photograph was taken after the successful maiden flight of the A-60 on 19 August 1981 or following the successful test of the LKAB on 27 April 1984. Almaz-Antey

The successes achieved with the LKO ground based laser weapon complexes in the early 1980's were due, in part, to what the Deputy General Designer at GSKB Almaz-Antey referred to in a 2014 interview as 'special multi-engined power drives, which provide an acceptable smooth operation and high precision with good accuracy correction drive support and guidance' (Kranaya Zvezda, 2014). The Deputy General Designer went on to detail other innovative areas of development ironed out with the LKO program, including the testing of a data system featuring 'high angular resolution' (Kranaya Zvezda, 2014). Experience gained in the LKO development was applied to the ALK program (Almaz-Antey).

Although western intelligence agencies were convinced that the Soviet Union was considerably ahead in development of laser weapon complexes (this was true in regard to ground based laser weapon complex experimentation) in 2014 a senior figure at Almaz-Antey stated that the USSR was slightly behind the United States in the development of an airborne laser complex. This is, in part, evinced by the fact that the United States recorded a successful test of a 400 kW CO_2 laser mounted on a converted Boing KC-135 (NKC-135) airborne laboratory on 23 February 1983, more than one full year before the first successful Soviet engagement of an airborne target by the LKAB carried onboard the A-60 1A1 ALK flying laboratory (Almaz-Antey). The Soviet Union, as noted above, had achieved considerably more success with its ground based experimental laser weapon complexes.

On 27 April 1984, A-60 1A1 took off for a development flight distinct from previous flights in that it would intercept and destroy a target in the air with its on-board laser complex. The aircraft took off from the base near Moscow.[16] Some seven hours and 18 minutes into the mission the go order was given for implementation of the test against an La-17K target aircraft that had taken off from a facility nearer the test area thousands of km distant from Moscow[17] (Almaz-Antey). The target was engaged and almost instantaneously a glow appeared in the area of the port wing structure, which was then engulfed in flames before breaking off, the target tumbling earthward to impact the surface.

The A-60 Ladoga 1A1 taxiing prior to or following a development flight in support of the Soviet Union's airborne laser weapon development program. Almaz-Antey

---

[16] This is thought to be Chkalovsky (Shchelkovo) airport near Moscow, but not confirmed

[17] It is unclear if the entire duration was calculated as airborne time. The test area was assumed to be near Sary Shagan in Soviet Kazakhstan

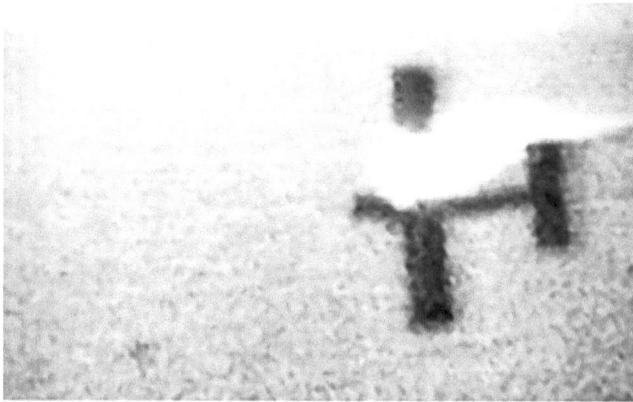

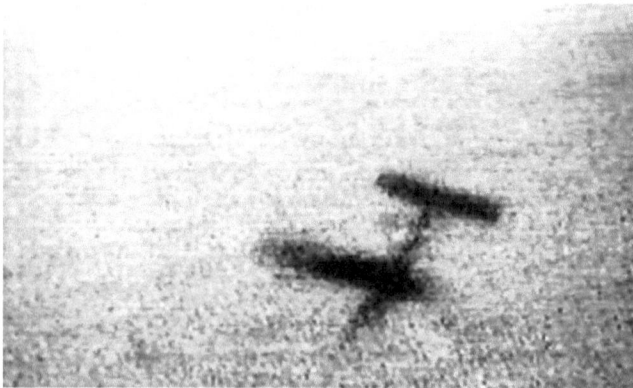

The Lavochkin La-17K target aircraft is struck on the inner port wing section by the thermal radiation beam originating from the A-60 1A1 during the 27 April 2984 live fire test. The damage and fire resulted in the wing breaking away from the aircraft which then tumbled to the ground out of control. Almaz-Antey

A-60 1A1 airborne laser flying laboratory during a test flight. Almaz-Antey

Research and development to further the Soviet laser complex programs continued into the 1990's. However, while there were advances, there were also setbacks, most notably with the loss of the A-60 1A1. This aircraft was widely reported in the West to have been destroyed in a fire resulting from a ground accident around mid-June 1986. While we know the aircraft was destroyed at some point it is unclear exactly what date this occurred. The burnt out wreck apparently remained at Chkalovsky (Shchelkovo) airport into the twenty first century.[18] A second A-60 flying laboratory with the index 1A2 was flown in 1991, this serving as the basis of the revamped airborne laser program, now conducted within the realm of the post-Soviet Russian Federation.

SEA BASED – As well as the ground based and airborne laser complex programs, both of which had a primary anti-airspace role, the Soviet Union was heavily invested in laser weapon development intended for application against space based threats and ground targets. One of the most oft quoted Soviet laser weapon developments concerned a maritime platform based laser system that is stated to have been installed on the Soviet Black Sea Fleet Auxiliary (oiler) *Dixon*. This weapon is oft stated to have been successfully tested against a shore based target. However, each telling of the *Dixon* tale is conveniently bereft of detail, which is replaced by hearsay or innuendo – nothing of substance that can be backed by government agency or design bureau documentation. If such a program was indeed tested in the Black Sea then it has to be assumed that it was conducted in very clear

---

[18] It has not been possible to locate this wreck on commercial satellite imagery dating back to 2003.

and calm conditions as even a moderately blustery sea state would have induced severe thermal blooming interference in the thermal radiation beams path to a target located onshore. Furthermore, the fact that the oiler *Dixon* was allocated to the Ukraine rather that the Russian Federation when the former Soviet Black Sea Fleet was divided in the early 1990's would suggest, strongly, that there was no laser complex equipment onboard at that time, if, indeed, any such equipment had ever been installed.

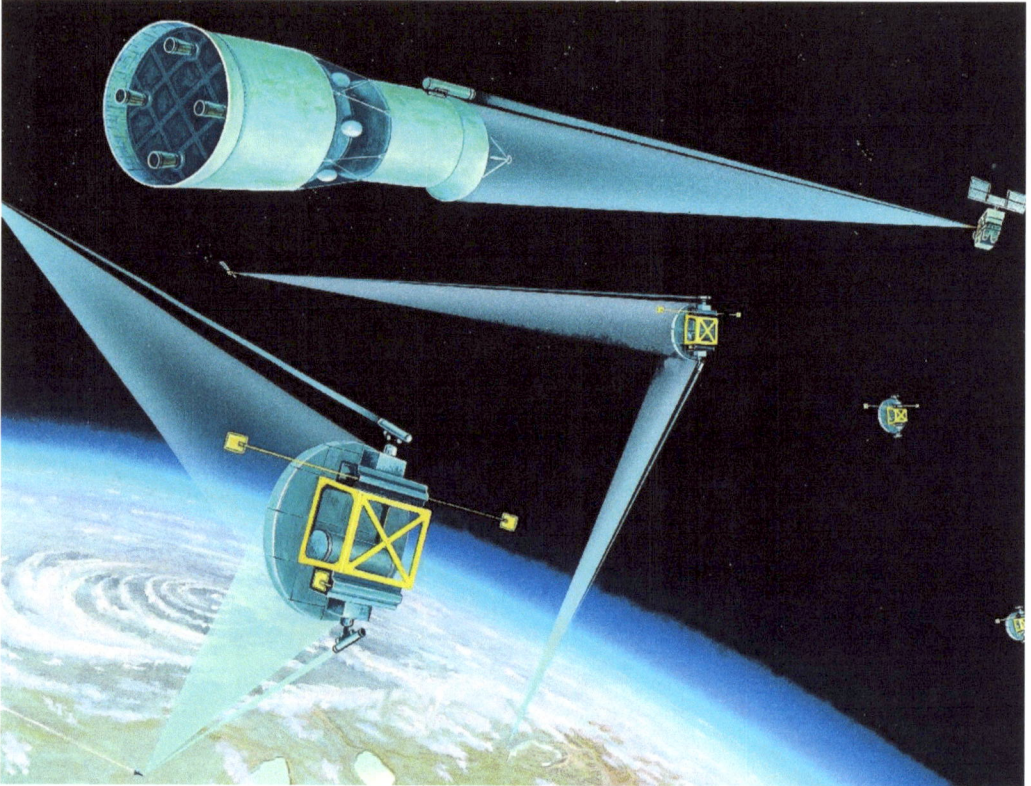

This DIA 1980's artwork depicts a Soviet space based equivalent to the United States Strategic Defense Initiative. The system is shown as orbital laser weapon stations and mirrors able to combat orbital space craft such as reconnaissance satellites and ground launched sub-orbital intercontinental ballistic missiles. DIA

ASAT – There is much more certainty in the Soviet laser armed ASAT (Anti-Satellite) program intended as a new generation ASAT to replace the Counter Space Defence Co-Orbital Satellite Fighter that had been operational since 1972.[19] We know with absolute certainty that this program existed through verifiable facts available from S.P. Korolev Rocket and Space Corporation Energia documentation.

---

[19] This ASAT capability, developed by OJSC 'Corporation' Space System Special, Comet, with input from many other bureau, organisations and institutions, was operational until 1993 (Harkins, 2017)

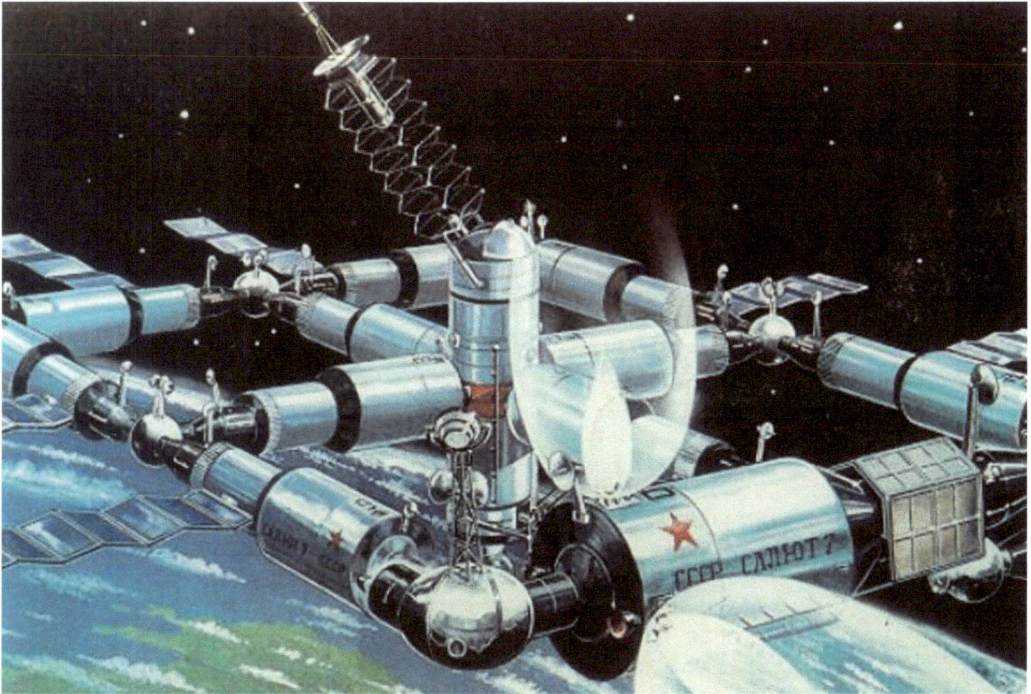

Resembling something from a 1980's movie set this DIA graphic depicts an early 1980's assessment of a then future Soviet orbital space station complete with a laser defence system that could threaten other spacecraft and possibly targets on Earth. While no such armed orbital stations were built, there were theoretical studies for such stations, particularly within Rocket and Space Corporation Energia. Above: The Buran reusable spaceplane, here landing following its only orbital flight on 15 November 1988, was to be employed to carry laser and missile armed ASAT vehicles into Earth orbit. DIA/Energia

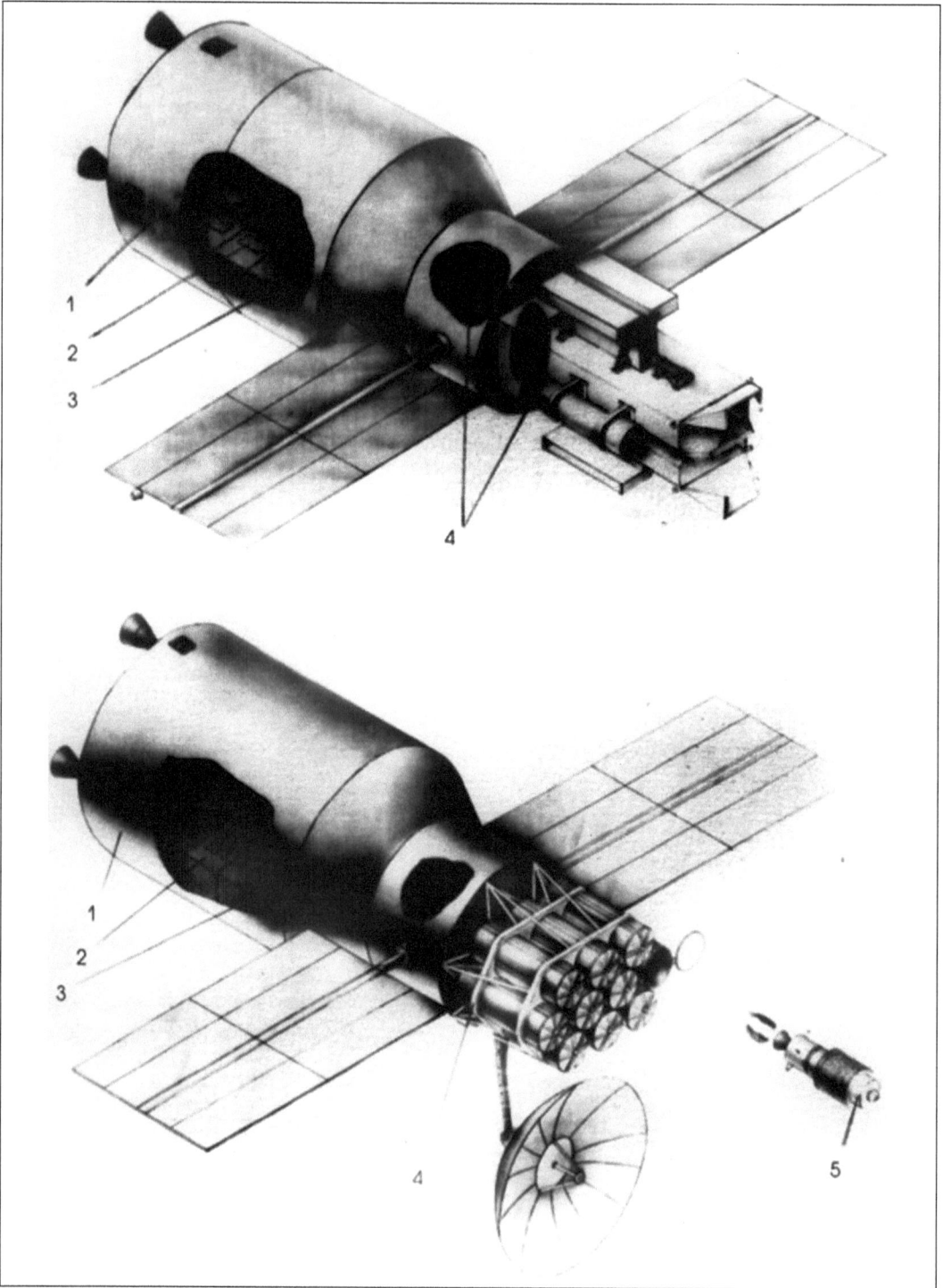

Spacecraft with laser weapon (top). 1. Aggregate compartment. 2. Instrument compartment. 3. Tanks ODU. 4. Onboard complex of special weapon. Spacecraft with missile armament (bottom). 1. Aggregate compartment. 2. Instrument compartment. 3 Tanks. 4. Onboard weapons complex. 5. Guided missile. Energia

The Co-Orbital Satellite Fighter had provided the Soviet Union with a first generation ASAT capable of intercepting spacecraft at orbital altitudes up to a few thousand km. The second generation ASAT was required to provide a capability against spacecraft orbiting at low medium and geostationary Earth orbits.[20] To achieve the medium and geostationary Earth orbit interception capability the ASAT spacecraft was to be equipped with a complex of missile weapons. The low Earth orbit ASAT capability was to be provided by a similar spacecraft armed with a laser weapon complex. The two types of craft were complementary and were intended to be operated in conjunction to provide an ASAT capability against the full spectrum of surveillance, communications and navigation satellite orbits (Energia). In the late 1970's and early 1980's the second generation ASAT program was being studied in conjunction with other space defence programs – the potential for destroying intermediate range and intercontinental range ballistic missiles and the potential for engaging high priority ground targets from space. These distinct, but related requirements were to be met through a family of space based platforms (Energia).

As noted above, for the ASAT role two separate spacecraft were to be developed, both based on the DOS-7K orbital space station. The ASAT derivatives of the DOS-7K were to be equipped with increased fuel capacity to feed the orbital maneuvering engines. The ASAT vehicles would be carried into Earth orbit as cargo aboard the Energia-Buran reusable space transport system (the Soviet equivalent of the US Space Shuttle) (Energia). However, as the ASAT vehicle was expected to be ready before the Energia-Buran system was cleared for operational deployment it was planned to use the Proton-K launch vehicle to carry development ASAT spacecraft into orbit (Energia & Khrunichev). Buran was also to be utilised to carry additional fuel into orbit for the refueling of the ASAT vehicles when required (Energia). It is unclear if such refueling would have concerned only the replenishment of the fuel tanks feeding the spacecraft engines or if it was also intended to provide additional fuel for the complex of laser weaponry. As the missile armed spacecraft was designed with a lower mass than that of the laser armed spacecraft, additional fuel for combat maneuvering was carried.

The orbital complex for engaging ground targets was also to be based on the DOS-7K orbital station. This would form a central hub for the docking of several additional autonomous modules. The autonomous combat modules, which were to be developed from an early iteration of the Buran reusable spacecraft would undock when the time came for combat deployment.[21] They would them move to an orbital point determined by the tasking against a specific target(s) on the surface of the Earth.

---

[20] Geostationary Earth orbit (also known as Geosynchronous orbit) = altitude of approximately 35790 km, which provides for the satellite to orbit Earth with a period the same as the period for a single rotation of the planet on its axis – 23 hours, 56 minutes, 4.09 seconds (this is known as the sidereal day)

[21] This would have been smaller in overall dimensions and overall mass compared to the Buran orbiter

Diagrammatic artwork depicting a design from studies for a Soviet space station able to strike Earth surface targets. 1. Transport ship 7K-SM. 2. Command module. 3 Base unit of the station. 4 Target module of the station. 5. Combat module (based on early iteration of the Buran reusable spaceplane (orbiter). Energia

TWENTY FIRST CENTURY – The economic downturn in the Russian Federation that followed the dissolution of the Soviet Union on 25 December 1991 continued through the rest of the decade. While many defence related research, development and procurement programs were cancelled outright, others were allowed to fall into a state of abeyance due to lack of state funding – in effect they existed in name only. The former Soviet ASAT (laser and missile armed) programs and, assuming its advancement was fact, the naval laser platform program were shelved, but the KLO and ALK programs were frozen, but not cancelled.[22] By keeping the KLO and ALK programs ticking over it was possible to preserve the Scientific/technical base that had been built up from the second half of the 1970's through 1991 (Almaz-Antey). Small-scale work on the ALK was able to continue through allocation of extra-budgetary funding. Through 1998, a much reduced in scale development program had achieved considerable success in the area of 'improving the characteristics of the main devices of laser systems' (Almaz-Antey). To generate additional funding in an era of sparse defence spending the laser research program, through cooperation between SSC TRINITI, D.V. Efremova and OAO NPO Almaz, advanced into other areas such as developing gas fire extinguishing laser complexes for the Russian Energy firm Gazprom (Almaz-Antey). Such programs were instrumental in the financial survival of the advanced laser development programs in the Russian Federation into the twenty first century. Going into the new century the Russian Federation state became reinvested in the KLO as part of its defence planning. This was energised by two major geopolitical and economic events: 1. The United States withdrawal from the ABM (Anti-Ballistic Missile) treaty on 13 June 2002 and the subsequent planning and later implementation of installations of American missile defence systems in Eastern Europe in an attempt to gain a strategic military advantage over the Russian Federation.[23] 2. A Russian economic revival that facilitated the resumption of state funding for such programs. The Russian KLO program, now reinvested with state funding, would be integrated into the wider plans to develop what is referred to as 'air and space defence' (Almaz-Antey). At this time the KLO and ALK programs shifted in direction – emphasis was now placed on countering enemy airborne platforms, in particular intelligence gathering assets, and spacecraft by disabling or

---

[22] It has not been possible to locate any documentation relating to cancellation of any laser weapon program with a primary naval function. Although this casts further doubt on the programs basis in fact, it is not conclusive as many documents remain classified

[23] The Anti-Ballistic Missile treaty, signed between the United States of America and the Union of Soviet Socialist Republics, had stood since 1972. The Russian Federation embarked upon a number of measures to nullify the effect of the US missile defence systems, including introduction of the Iskandar-M mobile short-range aero-ballistic/cruise missile complex designed to overcome missile defence systems by introducing a highly manoeuvrable vehicle able to circumvent interception (Harkins, 2016). Such systems have a counter missile defence strike role

[24] This would be inadequate for the intended role – to destroy ballistic missiles in the boost phase

otherwise interfering with optical sensors through thermal radiation in a process referred to as 'functional defeat' (Almaz-Antey).

Considering the goals of the respective airborne laser programs were fundamentally different, comparisons of the Russian ALK A-60 1A2 and the United States Boeing YAL-1, developed under the direction of the United States Missile Defence Agency, are of little value. As well as being intended to achieve fundamentally different strategic goals, funding for the American program dwarfed that of the Russian program, which is estimated to be around ten times below that afforded to the American program. The American system was designed as a hard kill weapon to destroy ballistic missiles in the boost phase through thermal overload at ranges of several hundreds of kilometers. A few successes were recorded – an obsolete R-17 derivative 'Scud' short/medium range ballistic missile was successfully intercepted at a range of 100 km in a test conducted in February 2010.[24] However, such tests, in near perfect conditions, belied the problems that still existed through thermal blooming, which would, following several failed tests, lead to the cancellation of the YAL-1 program. The difficulties of employing a laser as a hard kill weapon in a real world environment through the process of thermal overload at long ranges in Earth's atmosphere were omnipresent in the twenty first century just as they had been with the early laser weapon complexes in the late twentieth century.

The pace of development testing for the revamped Russian Federation ALK program ramped up in the second half of the second decade of the twenty first century. On 28 August 2009, a major milestone was passed when A-60 1A2 (AJISAI) employed new generation LKAB laser complex, pointed from the side of the aircraft, to target a spacecraft orbiting Earth at an altitude of ~1500 km. The laser was reflected from the aim point indicating a successful interception. This complex operation involved precise targeting of a thermal radiation beam from the airborne platform against a spacecraft orbiting at an altitude of 1500 km and travelling at hypersonic speed. To achieve this required a considerable research and development effort for the accurate detection and tracking of upwards of 20 spacecraft from which the target would be selected. The test phase for 'detection and angular tracking' of a spacecraft and 'angular guidance of the optical emission means of transport axes' – radiation beam – was stated to be 100% successful (Almaz-Antey).

While details of the ALK developments in the second decade of the twenty first century are notable for their almost complete absence, on 3 March 2018 the Russian Federation Government revealed that a mobile ground based complex of laser weapon – later named Peresvet – was in trials operation. Details of this new weapon complex are sparse, including its actual operational role(s) – to counter aircraft or spacecraft or both; a hard kill weapon or a soft kill weapon to disable air or space vehicle sensor optics. Any statement in favour of any of these roles or capabilities would, in late 2018, be purely speculative. However, it is clear that development of laser complexes as weapon systems continues in the Russian Federation as we approach the third decade of the twenty first century.

Previous page: While little is known about the Peresvet combat laser complex unveiled in March 2018, it can be stated that whilst the complex is mobile it is not intended to accompany a fielded army unless road integrity can be preserved. This is evinced by the fact that the low-loader element of the complex carrying the laser weapon is equipped with road wheels rather than high mobility-off road wheels. This page: The Peresvet laser complex in various stages of deployment. MODRF

GLOSSARY

ABM	Anti-Ballistic Missile (treaty)
AIM	Airborne Interception Missile
ALK	Airborne Laser Complex
ASAT	Anti-Satellite
AWACS	Airborne Warning and Control System
C	Centigrade
CDB	Central Design Bureau
CO_2	Carbon Dioxide
eV	Electron-volt – a unit of energy often used for specifying the energy of Photons. 1 eV = 1.602 x 10^{-19} Joules – electron charge
GPS	Global Positioning System (receives coordinates from orbiting satellite constellation)
ICBM	Intercontinental Ballistic Missile
IIR	Infrared/Imaging Infrared
Il	Ilyushin
IR	Infrared
IV	Roman Numeral number 4
KB-1	Experimental bureau formed in August 1973
kg	Kilograms
KLO	Complex of Laser Weapon
km	Kilometers
kW	Kilowatt
La	Lavochkin
Laser	Light Amplification by Stimulation of Radiation
LGB	Laser Guided Bomb
LKAB	Laser System Air Based
LL	Flying Laboratory
Mach	1 Mach – the speed of sound (this varies with altitude)
m s^{-1}	Metres per second
NATO	North Atlantic Treaty Organisation
NMUSAF	National Museum of the United States Air Force
RPG	Rocket Propelled Grenades
SDI	Space Defense Initiative
USSR	Union of Soviet Socialist Republics
Warsaw Pact	Formal Treaty of Friendship, Co-operation and Mutual Assistance between the USSR Socialist Republics and 7 Soviet Orbit states in Europe. This treaty, which took effect from 14 May 1955, was intended to counter the growing NATO alliance opposed to the Eastern Block
WMD	Optical-Mechanical Device
x	Tines – multiplication
=	Equal to
~	Approximately equal to (can also mean asymptotically equal)

CITATION GUIDE

Almaz-Antey	JSC Concern Aerospace Defence 'Almaz-Antey' (JSC GSKB Almaz-Antey)
Beriev	PJSC 'Beriev Aircraft' Experimental Design Bureau
Energia	S.P. Korolev Rocket and Space Corporation Energia
Krunichev	FSUE 'GKNPTs them. M.V. Khrunichev
MODRF	Ministry of Defence of the Russian Federation
Laspace	Scientific and Production Association, S.A. Lavochkin (Laspace)
Harkins, 2017	Harkins, H. (2017) *Counter Space Defence Co-Orbital Satellite Fighter: The Soviet Istrebitel Sputnik Anti-Satellite Complex*, Centurion Publishing, UK
Harkins, 2016	Harkins, H. (2016) *Iskander, Mobile Tactical Aero-Ballistic/Cruise Missile Complex*, Centurion Publishing, UK
Green & Jones, 2015	Green, S.F. & Jones, M.H. (2015) *An Introduction to the Sun and Stars*, Second Edition, Cambridge University Press
NMUSAF	National Museum of the United States Air Force
CIA	Central Intelligence Agency
DIA	Defence Intelligence Agency
WP	Washington Post, 6 September 1985 (reproduced in CIA-RDP90-00965R000100120046-1, approved for release 2012-01-10)
Kranaya Zvezda	Kranaya Zvezda News

ABOUT THE AUTHOR

Hugh Harkins FRAS is a historian and author with an extensive research background in astro/geophysics and studies/research in the wider scientific, aeronautic, astronautic and nautical technical and historical fields. He is also involved in research in the field of Scottish history, which formed a significant element of an otherwise scientific undergraduate degree. Hugh has published in excess of sixty books; non-fiction and fiction, writing under his given name as well as utilising several pseudonyms. He has also written for several international magazines, whilst his work has been used as reference for many other projects ranging from the aviation industry, international news corporations and film media to encyclopaedias, museum exhibits and the computer gaming industry. Hugh is a member of the Institute of Physics and is an elected Fellow of the Royal Astronomical Society. He currently resides in his native Scotland. Other titles by the author include:

Russia's Coastal Missile Shield - Bal-E & Bastion Mobile Coastal Cruise Missile Complexes
Iskander - Mobile Tactical Aero-Ballistic/Cruise Missile Complex
Orbital/Fractional Orbit Bombardment System - The Soviet Globalnaya Raketa
Counter-Space Defence Co-Orbital Satellite Fighter
Russia's Strategic Missile Carrier/Bomber Roadmap 2018-2040 – PAK DA, Tu-160M2, Tu-95MSM & Tu-22M3M
Sukhoi T-50/PAK FA - Russia's 5th Generation 'Stealth' Fighter
Sukhoi Su-35S 'Flanker' E - Russia's 4++ Generation Super-Manoeuvrability Fighter
Sukhoi Su-34 'Fullback'
Sukhoi Su-30MKK/MK2/M2 - Russo Kitashiy Striker from Amur
Soviet Mixed Power Experimental Fighter Aircraft – Piston-Liquid Propellant Rocket Engine/Piston-Ramjet/Piston-Pulsejet & Piston-Compressor Jet Engine Designs of the 1940's
MiG-35/D 'Fulcrum' F – Towards the Fifth Generation
Air War over Syria, Tu-160, Tu-95MS & Tu-22M3 - Cruise Missile and Bombing Strikes on Syria, November 2015-February 2016
Sukhoi Su-27SM(3)/SKM
Russian/Soviet Aircraft Carrier & Carrier Aviation Design & Evolution Volume 1 - Seaplane Carriers, Project 71/72, Graf Zeppelin, Project 1123 ASW Cruiser & Project 1143-1143.4 Heavy Aircraft Carrying Cruiser
Light Battle Cruisers and the Second Battle of Heligoland Bight
British Battlecruisers of World War 1 - Operational Log, July 1914-June 1915
Eurofighter Typhoon - Storm over Europe
North American F-108 Rapier - Mach 3 Interceptor
Convair YB-60 - Fort Worth Overcast
Boeing X-36 Tailless Agility Flight Research Aircraft
X-32 - The Boeing Joint Strike Fighter
X-35 - Progenitor to the F-35 Lightning II
X-45 Uninhabited Combat Air Vehicle
Into The Cauldron - The Lancaster MK.I Daylight Raid on Augsburg
Hurricane IIB Combat Log - 151 Wing RAF, North Russia 1941
RAF Meteor Jet Fighters in World War II, an Operational Log
Typhoon IA/B Combat Log - Operation Jubilee, August 1942
Defiant MK.I Combat Log - Fighter Command, May-September 1940
Blenheim MK.IF Combat Log - Fighter Command Day Fighter Sweeps/Night Interceptions, September 1939 - June 1940
Fortress MK.I Combat Log - Bomber Command High Altitude Bombing Operations, July-September 1941